ANIMAL RHYME TIME!

RHYME TIME WITH SHARKS!

BY JONAS EDWARDS

Gareth Stevens PUBLISHING

Please visit our website, www.garethstevens.com. For a free color catalog of all our high-quality books, call toll free 1-800-542-2595 or fax 1-877-542-2596.

Cataloging-in-Publication Data

Names: Edwards, Jonas.
Title: Rhyme time with sharks! / Jonas Edwards.
Description: New York : Gareth Stevens Publishing, 2021. | Series: Animal rhyme time! | Includes glossary and index.
Identifiers: ISBN 9781538256145 (pbk.) | ISBN 9781538256169 (library bound) | ISBN 9781538256152 (6 pack)
Subjects: LCSH: Sharks–Juvenile literature.
Classification: LCC QL638.9 E39 2021 | DDC 597.3–dc23

First Edition

Published in 2021 by
Gareth Stevens Publishing
111 East 14th Street, Suite 349
New York, NY 10003

Editor: Kate Mikoley

Photo credits: Cover, p. 1 Andrea Izzotti/Shutterstock.com; cover, pp. 3–24 (music notes) StockSmartStart/Shutterstock.com; p. 5 Alastair Pollock Photography/Moment/Getty Images; p. 7 © Andrey Nekrasov / Barcroft Media via Getty Images; p. 9 Richard Robinson/Cultura/ Getty Images; p. 11 Seashell World/Shutterstock.com; p. 13 Brad Leue / Barcroft Media via Getty Images / Barcroft Media via Getty Images; p. 15 Jeff Rotman/ The Image Bank / Getty Images Plus; p. 17 Image Source/Getty Images; p. 19 Barbara Ash/Shutterstock.com; p. 21 Benjamin Lowy/ The Image Bank / Getty Images Plus.

Printed in the United States of America

CPSIA compliance information: Batch #CS20GS: For further information contact Gareth Stevens, New York, New York at 1-800-542-2595.

CONTENTS

Boldface words appear in the glossary.

Big Fish in the Sea

Hello! I'm a whale shark. Do you see me?
I'm the biggest fish in the sea!
I'm as long as a school bus.
Whale sharks are calm and don't make a fuss.
Not all sharks are like me.
Come, read along and you'll see.

Shark Snacks

Sharks are a kind of **predatory** fish.
Smaller fish are a favorite dish.
Sharks sneak up on **prey** and
may eat them whole!
As part of the ocean **food chain**,
sharks play a top part, or role.

Slow or Speedy?

Hundreds of kinds of sharks live in
the ocean.
Some move so slowly, it's like they're
barely in **motion**.
Others are fast, with bodies made
for speed.
This helps them surprise and catch
the food that they need.

MAKO SHARK

Shark Bones

Sharks have **skeletons**, but no bones.
This is true even when they're old
and fully grown.
Their skeletons are made of
cartilage, in fact.
This is the same matter that keeps
your nose **intact**.

SHARK JAW

Sharp Teeth

Sharks are known for their teeth,
sharp and scary.
But how sharp and big they are
sure can **vary**.
One thing's for sure about teeth—they
have many.
They fall out, but the tooth fairy won't
give them a penny!

When a shark loses a tooth
it's no big deal.
Rows of more are ready to help
with the next meal.
If one falls out, a new one will take
its place.
In just a few weeks it'll move
to fill the space.

Great White Sharks

One famous shark is called a
great white.
They're large, fast, and have a
strong bite.
Their teeth can cut prey's skin and
break bone.
I hope you never see one, they're
best left alone!

Give Us Space

Lots of sharks can **attack** and are scary.
If you ever see one it's best to be **wary**.
Some sharks just want you to let
them be.
They'd say, "I won't hurt you if you
don't hurt me."

LEOPARD SHARK

Don't Mess with Us!

Now you know not to mess with
a shark.
If you want to see one, try a zoo
or animal park.
You don't want to come across one
at sea.
They're important but dangerous. I bet
you'd agree!

GLOSSARY

attack: to try to harm someone or something

cartilage: tough, bendable tissue that makes up part or all of a skeleton

food chain: a series of organisms in which one uses the next lowest one as food

intact: not broken

motion: movement

predatory: living by hunting and eating other animals

prey: an animal hunted by other animals for food

skeleton: the bony frame of the body

vary: to be different or to become different

wary: not having complete trust in something that is or could be dangerous

FOR MORE INFORMATION

BOOKS

Moses, Brian. *Animal Poems*. New York, NY: Windmill Books, 2018.

Potts, Nikki. *Shark Moves Out of the Ocean*. North Mankato, MN: Picture Window Books, 2020.

West, David. *Sharks and Other Fish*. New York, NY: Windmill Books, 2018.

WEBSITES

Great White Shark
kids.nationalgeographic.com/animals/fish/great-white-shark/
Learn even more about great white sharks here.

Leopard Shark
kids.sandiegozoo.org/animals/leopard-shark
Read all about leopard sharks on this website from the San Diego Zoo.

Sharks
www.dkfindout.com/us/animals-and-nature/fish/sharks
Find out more about sharks on this interactive page.

INDEX